UNDERSTANDING SCIENCE

TEMPERATURE

PUBLISHED BY SMART APPLE MEDIA
1980 Lookout Drive, North Mankato, Minnesota 56003

Copyright © 2003 Smart Apple Media. International copyright reserved in all countries. No part of this book may be reproduced in any form without written permission from the publisher.

PHOTOGRAPHS by Richard Cummins, The Image Finders (Jim Baron, Lois Wheeler), Sally Myers, Tom Myers, NASA, Photri-Microstock, Tom Stack & Associates (Gary Milburn, SAO/IBM Research/TSADO, Therisa Stack, Spencer Swanger, Greg Vaughn), Unicorn Photography (Rod Furgason, Ed Harp, Dick Young)

DESIGN AND PRODUCTION Evansday Design

LIBRARY OF CONGRESS CATALOGING-IN-PUBLICATION DATA
Frisch, Joy.
Temperature / by Joy Frisch.
p. cm. — (Understanding science)
Includes index.
Summary: Introduces temperature, how it is measured, its effect on Earth, and such future concerns as global warming. Includes directions for making a thermometer.
ISBN 1-58340-159-8
1. Temperature—Juvenile literature. [1. Temperature.] I. Title. II. Understanding science (North Mankato, Minn.).
QC271.4.F75 2002
536'.5—dc21   2001054926

First Edition

9 8 7 6 5 4 3 2 1

Temperature
UNDERSTANDING SCIENCE

[Joy Frisch]

INTRODUCTION**TEMPERATURE**

What is the temperature outside today? Temperature affects our lives every day. It helps us decide which clothes we will wear. Our outdoor activities are determined by the temperature outside. The sun's rays heat up the earth and create the weather around us. If it were not for the warmth from the sun, no human beings, animals, or plants would be able to live on earth.

THE SCIENCE OF TEMPERATURE

Temperature is a measure of how hot or cold something is. Heating something makes its temperature rise. When it cools off, its temperature falls. All **matter** is made of tiny particles called **molecules** that are always moving. Hot molecules move faster than cold ones, so the more an object or organism is heated, the faster its molecules move around. When molecules are moving, they create **kinetic energy**. Temperature is our way of measuring the **energy** that the earth has absorbed from the sun's rays. Many substances change their appearance when heated or cooled. Depending on the temperature, some forms of matter can be a solid, liquid, or gas. For instance, water, when heated, will evaporate and turn to steam, or water vapor. If it is cooled enough, water will turn to ice. Many substances get

Matter is anything in liquid, solid, or gas form that has weight and takes up space.

Molecules are the smallest parts of matter.

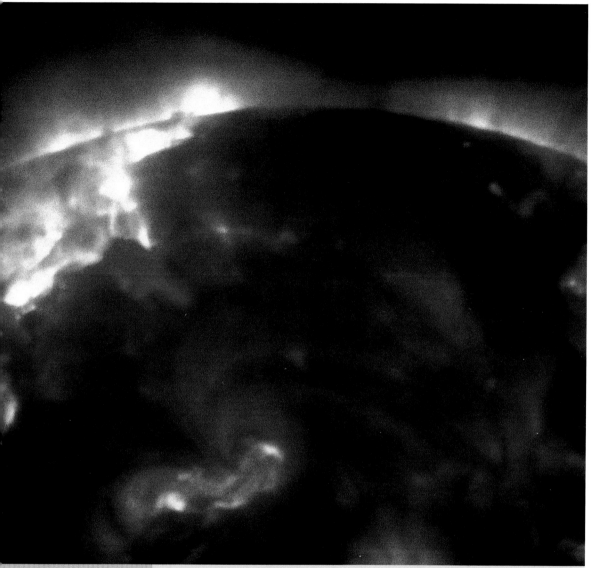

THE SURFACE OF THE SUN IS ABOUT 10,000 °F (5,537 °C)

When railroad tracks are built, small gaps are left in the metal tracks. The gaps allow the tracks room to expand on hot days. Without them, the tracks would buckle and break.

Kinetic energy *is energy associated with motion.*

DEPENDING ON TEMPERATURES, WATER CAN BE LIQUID, ICE, OR VAPOR

bigger, or expand, when they are heated, and get smaller, or contract, when they are cooled. Liquids usually expand more than solids do when they are heated, and gases expand most of all. For instance, heating a kettle of hot chocolate on a stove will cause it to boil and expand. It will boil over if unattended. Liquid that evaporates will spread out in the air and seem to disappear.

Energy *is usable power; the ability to do work.*

Hot-air balloons float through the air because the hot air inside the balloon is warmer than the air outside the balloon. The heated air expands and gets lighter, making the balloon rise.

HEAT FROM THE SUN

The sun is like a giant furnace. It is a ball of burning gases whose temperature at the surface is about 10,000 °F (5,537 °C). The temperature deep inside the sun is about 25 million °F (13.9 million °C). Without the sun, there would be no heat to warm our planet. Every type of weather occurs because the heat of the sun keeps the air in Earth's **atmosphere** constantly in motion. The sun's position relative to the earth helps determine the temperature of an area. When the sun is high in the sky, its rays strike the ground directly, and its heat is strong. When the sun is low in the sky, its rays strike the ground at an angle, and its heat is spread out and weaker. Because of these variations, some parts of the world get hot weather and some places get cold weather on the same day. Not all places on the earth's surface receive the same amount of heat from the sun. The hottest spots are located near the **equator**, and the

*The **atmosphere** is the layer of gases, mainly oxygen and nitrogen, that surrounds the earth.*

STRONG HEAT AND LITTLE RAINFALL CREATE DESERT CLIMATES

To help open a jar, place it under hot running water. The heat makes the metal lid expand so that it squeezes less against the jar and is easier to turn.

*The **equator** is the imaginary line around Earth; it divides the northern hemisphere from the southern hemisphere.*

*An **axis** is an imaginary line upon which a rotating body, such as the earth, turns.*

coldest spots are in the Arctic Circle and Antarctica. The unequal heating of Earth sets the air in the atmosphere in motion. Cold air is heavier than warm air, which makes it sink. Lighter, warm air rises above cool air. Air near the equator is heated, becomes lighter, and rises. At the poles, colder and heavier air settles downward and moves in to take the place of the warm air that has risen. These rising and sinking air masses create all wind and weather patterns. The earth is tilted at an angle on its **axis**, and its surface is curved. This is why the sun's rays strike some areas directly and other areas at an angle. The change in temperature due to this effect causes the **seasons**. Over most of the earth, temperatures change with the seasons. As one part of the earth tilts away from the sun, the air gets colder, and it becomes winter. In the part of the earth that tilts toward the sun, the air gets warmer, and it be-

Seasons *are the different times of year when weather and temperatures change, including winter, spring, summer, and fall.*

comes summer. Near the equator, the seasons do not differ much in temperature. The sun's rays strike almost directly year round, so the temperature remains high. The further away from the equator a place is, the lower its summer and winter temperatures and the larger the difference between the two. **Solar radiation** is the main cause of temperature differences. That is, temperatures increase during daylight hours when the sun is out and decrease at night.

ANTARCTICA RECEIVES LITTLE DIRECT SUNLIGHT, EVEN IN THE SUMMER

Solar radiation *is heat and light energy from the sun.*

ABSORPTION AND REFLECTION

Different surfaces of the earth absorb different amounts of the sun's energy. **Surface temperature** differs according to the kind of land cover. Forests, sand, and bare soil absorb the largest amounts of the sun's energy. A tar road feels very hot on a warm summer day because it absorbs a lot of the sun's energy. Clouds also absorb heat well and can soak up more than 80 percent of the sun's heat on a cloudy day. Snow and ice, however, reflect as much as 90 percent of the sun's energy back into the atmosphere. The ocean also reflects most of the sun's energy. About one-third of the energy that reaches Earth's atmosphere is reflected back into space. The remaining two-thirds is absorbed. Sunlight heats the ground, which in turn warms the air near the surface, and the atmosphere prevents most of the heat from escaping into space. This

Surface temperature is the amount of radiation or heat that is absorbed by different surfaces.

DENSE CLOUD COVER HAS A COOLING EFFECT ON THE EARTH

A city with many buildings and roads tends to be warmer at night than its surroundings. Building materials such as bricks, stone, and concrete absorb a lot of heat during the day, then slowly release it during the night.

HIGH MOUNTAIN PEAKS MAY BE COVERED BY SNOW IN ALL SEASONS

method of trapping and holding heat is known as the **greenhouse effect**. Changes in altitude also contribute to differences in temperature. Temperatures drop at high altitudes, where the air is thin and therefore has fewer molecules to heat. Very high mountain tops remain snow-covered during summer months, even while flowers bloom on the lower slopes. Mountain elevations are cold because air temperatures drop a few degrees with each 1,000 feet (305 m) of altitude. On high peaks, temperatures may remain below freezing year round.

*The **greenhouse effect** is the heating effect caused by gases in the atmosphere trapping heat from the earth's surface.*

Trees are impacted by temperature differences. Each ring in a cut tree trunk shows one year's growth. If the ring is wide, the tree grew well and the weather was warm. If the ring is narrow, then the weather was cold.

MEASURING TEMPERATURE

The scientific study of weather, called meteorology, began in Italy. In 1593, Galileo Galilei, a mathematician and philosopher, invented the first **thermometer** to measure changes in temperature. He put liquid alcohol inside a glass bulb the size of a chicken egg with a clear, thin tube sticking out of it. He and his friends then watched the liquid rise and fall in the tube as the temperature changed. The first mercury thermometer, the kind still used today, was developed in 1724 by German physicist Gabriel Fahrenheit. Mercury is a silver-colored metal that is liquid at room temperature. Thermometers work because liquids expand more than solids as their temperatures increase. A glass thermometer contains a narrow, hollow tube with a base that is filled with a liquid such as alcohol or mercury. When the temperature rises, the liquid expands and moves up the tube. When the air tem-

*A **thermometer** is an instrument used to measure the temperature of air or water.*

LIGHT COLORS HELP TO REDUCE HEAT BY REFLECTING SUNLIGHT

In countries with a hot, sunny climate, buildings are often painted white to reflect the heat. People also wear light-colored clothing to stay cool.

perature drops, the liquid cools off, takes up less room, and slides back down the tube. Meteorologists, scientists who study weather, always measure temperature in the shade. They do this to get accurate readings of the air temperature. A temperature reading taken in direct sunlight would be higher and more inconsistent than one taken in the shade. After creating his thermometer, Fahrenheit developed a scale of measurement in which water boils at 212 degrees and freezes at 32 degrees. Temperatures measured on this scale are designated as degrees Fahrenheit (F). In 1742,

The coldest climate in the world is in Antarctica. The average temperature there is −69.8 °F (−56.6 °C). The lowest air temperature ever recorded occurred there when the temperature reached −128.6 °F (−89.2 °C) in 1983.

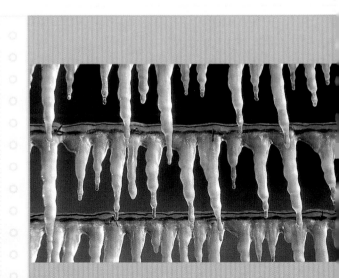

TEMPERATURES CAN INCREASE QUICKLY AS THE SUN RISES IN THE SKY

Swedish astronomer Anders Celsius developed a scale in which the freezing point of water is zero degrees and the boiling point is 100 degrees. Temperatures measured on this scale are designated as degrees Celsius (C). The Fahrenheit scale is used in the United States; the Celsius scale is used in much of the rest of the world, including Europe and Canada. The lines on a thermometer show the temperature in degrees. Marked on one side is the Fahrenheit scale. Marked on the other side is the Celsius scale.

KEEPING COOL, STAYING WARM

People live in different **climates** around the world, from the polar areas to the equator. In order to live comfortably, particularly where the climate is harsh, people design houses, clothes, and lifestyles to fit the conditions of their particular climate. All of our outdoor activities are affected by temperature. On a winter day, a person is likely to wear warm clothes while skiing in the snow. On a hot and humid day, swimming at a beach or sitting in the shade of a tree will help to keep a person cool. On cold days, the wind can make the air temperature feel colder than it actually is. This is called the windchill factor. The temperature of a healthy human body is 98.6 °F (37 °C). The body reacts to different temperatures in various ways to help it adjust its own temperature. For example, when a person gets very hot, extra body heat is released by

Climates *are the average weather conditions experienced in an area over a long period of time.*

A CULTURE'S CLOTHING IS LARGELY DETERMINED BY TEMPERATURE

The South Pole has fewer hours of sunshine per year than any other place in the world. During the winter there, the sun does not rise for 182 straight days due to the tilt of the earth.

A PERSON'S BODY IS COOLED AS WATER EVAPORATES FROM THE SKIN

allowing water to evaporate from the skin, which is called sweating. When it is very cold, a person's blood vessels constrict, or become narrower, letting less blood through and minimizing heat loss from the body. Blood vessels are our bodies' temperature controls. They widen when we are warm because molecules expand when heated. They constrict when we are cold to conserve heat. In extremely cold conditions, a person's fingers and toes may feel numb because the flow of blood to those parts is lessened. It is important to consider the color of a person's clothing when spending time outdoors. Dark-colored clothing is better at absorbing heat radiation from the sun than light clothing. A person wearing a black t-shirt will feel warmer than a person wearing a white t-shirt. Light-colored and shiny surfaces reflect the sun's rays and reduce the amount of heat that is absorbed.

Like many small creatures, grasshoppers are sensitive to changes in the weather. They chirp louder and louder as the temperature rises.

EARTH'S TEMPERATURE

In recent years, scientists have become concerned about the effects of human activities on the world's weather. Most meteorologists believe that average temperatures around the world are gradually rising due to increased greenhouse gases, which absorb heat from the sun. They believe that the earth is getting warmer because there are too many greenhouse gases building up in the atmosphere. This effect is known as **global warming**. Greenhouse gases are beneficial in the right quantities; like the panes of glass in a greenhouse, they trap the sun's heat and keep the earth warm. But many scientists fear they are now keeping the earth too warm. The extra greenhouse gases come from pollution. Carbon dioxide is the main greenhouse gas. The most harmful gases come from the fumes of cars, forest fires, and the burning of **fossil fuels**

Global warming *is the increase in world temperatures caused by heat being trapped by greenhouse gases.*

MANY SCIENTISTS THINK THAT POLLUTION IS WARMING THE PLANET

The hottest place in the world is Dallol, Ethiopia, in East Africa, where temperatures average 94 °F (34.4 °C). The highest temperature ever recorded occurred in Libya, on the edge of the Sahara Desert, in 1922. The temperature reached 136.4 °F (58 °C).

Fossil fuels *are organic materials from the earth, such as oil, coal, and natural gas, which can be burned to produce energy.*

in factories. Cars give out all kinds of pollutants, but especially carbon dioxide. If world temperatures rise, world climates and the lives of people and wildlife will be greatly affected. Some areas may flood, while other areas may turn into deserts, forcing people and animals to find new places to live. The weather may also change dramatically, causing more powerful storms. The sun heats the earth to keep us warm and help plants grow. But the temperature of the world is affected by human activities as well. Too much pollution may be increasing worldwide temperatures. To

High in the atmosphere, air temperatures are always below freezing, and the wings of airplanes can easily become coated with ice. All jetliners must have de-icing equipment to keep their wings from freezing.

THE WEATHER COULD CHANGE DRAMATICALLY IF EARTH GETS TOO HOT

prevent global warming, people need to work together to minimize pollution. By doing this, we can help ensure that our home planet remains alive and well at a healthy temperature.

THERMOMETERS LET US KNOW WHEN TEMPERATURES ARE DANGEROUS

TEMPERATURE EXPERIMENT

To measure temperature, you can make your own thermometer. The liquid you use in the thermometer will react to different levels of heat, thereby revealing different temperatures in changing conditions.

WHAT YOU NEED

Water

Rubbing alcohol

A clear, narrow-necked plastic bottle

Food coloring

A clear plastic drinking straw

Modeling clay

WHAT YOU DO

1. Fill the bottle about one-eighth full with equal parts water and rubbing alcohol.
2. Add a few drops of food coloring and mix it with the water and alcohol.
3. Put the straw in the bottle. The end should dip into the liquid, but not touch the bottom.
4. Seal the neck of the bottle with the modeling clay so that the straw stands upright. The straw should be the only way air can get in or out of the bottle. Set the bottle outside on a warm day for half an hour. Then set the bottle inside a refrigerator for half an hour.

WHAT YOU SEE

What happened when you set your thermometer outside in the sun? What happened in the refrigerator? Like the mercury in a store-bought thermometer, the alcohol in your thermometer expands as it warms up. The expanding alcohol no longer fits into the bottom of the bottle and moves up the straw. If your thermometer gets too hot, the liquid will pour out the top of the straw.

INDEX

A
air, 9, 12-13, 17
altitude, 17, 28

B
blood vessels, 24-25

C
Celsius, Anders, 21
Celsius measurement scale, 21
climates, 22, 28
colors, 19, 25

E
equator, 10, 12, 13

F
Fahrenheit, Gabriel, 18, 20
Fahrenheit measurement scale, 20, 21

G
Galilei, Galileo, 18
global warming, 26
greenhouse effect, 17, 26

H
heat absorption, 14, 15, 25, 26

K
kinetic energy, 6, 8

M
matter, temperature's effect on, 6, 7, 9, 11, 14, 15, 17, 18, 20, 21, 25, 31
meteorology, 18, 20, 26
molecules, 6, 17, 24

P
pollution, 26, 28-29

S
seasons, 12, 13
solar radiation, 13
sun, the, 10, 12, 13, 14

T
temperature, 6, 10, 12, 14, 20, 21, 22, 27
   coldest air, 20
   definition of, 6
   human body, 22
   measuring, 20, 21
   sun, 10
   surface, 10, 12, 14
   warmest air, 27
thermometers, 18, 20, 21, 31
   mercury, 18

W
weather, 10, 12, 13, 20, 26, 28
wind, 12
windchill factor, 22